Breaking Boundaries Students Embrace International Education Technology Integration

Almanzo

Copyright © [2023]

Title: Breaking Boundaries Students Embrace International Education Technology Integration
Author's Almanzo

This book was printed and published by [Publisher's: **Almanzo**] in [2023]

ISBN:

TABLE OF CONTENT

Chapter 1: Introduction to International Education Technology Integration

The Importance of International Education Technology Integration

The Benefits of International Education Technology Integration

The Global Impact of International Education Technology Integration

Chapter 2: Understanding Comparative Education Systems

Exploring Different Education Systems around the World

Comparing Educational Approaches and Methods

Examining Challenges and Successes in Education Systems

Chapter 3: The Role of Technology in International Education

Integrating Technology in the Classroom

Enhancing Learning through Educational Technology

Overcoming Barriers to Technology Integration in Education

Chapter 4: Embracing International Education Technology Integration

Fostering Intercultural Communication and Collaboration

Developing Global Competencies through Technology

Building Digital Literacy Skills for the Global Age

Chapter 5: Case Studies of International Education Technology Integration

Successful Examples of Technology Integration in Schools

Impact of Technology on Student Engagement and Learning Outcomes

Lessons Learned from International Education Technology Integration

Chapter 6: Empowering Students in the Digital Age

Leveraging Technology for Personalized Learning

Promoting Creativity and Innovation through Technology

Preparing Students for Global Careers and Opportunities

Chapter 7: Overcoming Challenges in International Education Technology Integration   43

Addressing Infrastructure and Access Issues

Training and Professional Development for Educators

Navigating Ethical and Privacy Concerns in Technology Integration

Chapter 8: Future Trends in International Education Technology Integration   49

Emerging Technologies in Education

The Role of Artificial Intelligence in Education

Envisioning the Future of International Education Technology Integration

Chapter 9: Conclusion   55

Recap of Key Findings and Takeaways

Empowering Students to Embrace International Education Technology Integration

Chapter 1: Introduction to International Education Technology Integration

The Importance of International Education Technology Integration

In today's interconnected world, the importance of international education technology integration cannot be overstated. As students in the field of Comparative and International Education, it is crucial for us to understand the significance of incorporating technology into our learning experiences. This subchapter aims to shed light on the various reasons why international education technology integration is essential for our growth and success.

First and foremost, integrating technology into our education allows us to break through geographical boundaries. Through online platforms, virtual classrooms, and collaborative tools, we can connect with students and educators from around the world. This global interaction provides us with the opportunity to gain diverse perspectives, learn about different cultures, and broaden our understanding of international issues. By embracing technology, we can transcend physical limitations and engage in meaningful cross-cultural dialogues.

Furthermore, international education technology integration equips us with the necessary skills to thrive in the 21st century workforce. In an increasingly digital world, employers seek individuals who are adept at utilizing technology to solve complex problems, collaborate remotely, and adapt to rapidly changing environments. By familiarizing ourselves with various educational technologies, such as online research tools, video conferencing platforms, and data analysis software, we enhance our marketability and increase our chances of professional success.

Moreover, international education technology integration promotes self-directed learning and enhances our critical thinking abilities. With access to a vast array of online resources, we can delve deeper into topics of interest, explore alternative perspectives, and develop our analytical skills. Additionally, technology facilitates personalized learning experiences, allowing us to tailor our education to our individual needs and learning styles. This empowers us to take ownership of our education and develop a lifelong love for learning.

Lastly, international education technology integration fosters global citizenship and empathy. Through online collaborations, virtual exchange programs, and digital storytelling, we can connect with individuals from diverse backgrounds and build meaningful relationships. This not only enhances our intercultural competence but also promotes mutual understanding, tolerance, and empathy. By utilizing technology as a tool for global connection, we can become active participants in creating a more inclusive and interconnected world.

In conclusion, international education technology integration plays a pivotal role in our growth and development as students in the field of Comparative and International Education. By breaking through geographical boundaries, equipping us with essential skills, fostering self-directed learning, and promoting global citizenship, technology integration opens up a world of possibilities for our education. Embracing technology as a tool for international collaboration and learning will undoubtedly shape us into well-rounded individuals, prepared to navigate the challenges and opportunities of an increasingly globalized society.

The Benefits of International Education Technology Integration

In today's interconnected world, education is no longer confined to the boundaries of a single country. With the advent of technology, students now have the opportunity to embrace international education and expand their horizons like never before. This subchapter explores the numerous benefits of international education technology integration, shedding light on how it can enrich the learning experience for students across the globe.

First and foremost, integrating technology into international education allows students to gain a deeper understanding of different cultures and societies. Through virtual exchange programs, online collaborations, and video conferences, students can connect with peers from diverse backgrounds, fostering a sense of global citizenship and empathy. By interacting with students from different countries, they can gain insight into different perspectives, values, and traditions, broadening their worldview.

Furthermore, international education technology integration offers students the opportunity to develop crucial 21st-century skills. As they navigate through virtual platforms and collaborate with peers from different time zones, students enhance their digital literacy, communication, and problem-solving abilities. These skills are highly sought after in today's global job market, giving students a competitive edge and opening doors to exciting career opportunities worldwide.

Additionally, technology integration in international education promotes language learning and cultural exchange. Through language learning apps, online language exchange platforms, and virtual classrooms, students can engage in real-time conversations with native speakers. This not only enhances their language proficiency but also provides a platform to learn about different cultures, traditions, and customs. Such interactions foster mutual understanding and respect, breaking down barriers and promoting intercultural dialogue.

Moreover, international education technology integration allows students to access a vast pool of educational resources from around the world. Digital libraries, open educational resources, and online courses provide students with a wealth of information, enabling them to explore diverse subjects and pursue their interests. This democratization of knowledge empowers students, irrespective of their geographical location, to access high-quality education and engage in lifelong learning.

In conclusion, the integration of technology in international education brings with it a multitude of benefits for students. From fostering global citizenship to developing essential skills, promoting cultural exchange to providing access to a wealth of educational resources, the possibilities are endless. By embracing international education technology integration, students can break boundaries and embrace a world of learning opportunities, preparing themselves for success in an increasingly interconnected global society.

The Global Impact of International Education Technology Integration

In today's interconnected world, the integration of education technology has become increasingly crucial for students seeking a competitive edge. The rapid advancements in technology have revolutionized the way we learn and communicate, breaking down the barriers of traditional education systems. This subchapter explores the global impact of international education technology integration, highlighting its significance for students in the field of comparative and international education.

International education technology integration refers to the incorporation of technology tools and platforms into educational practices on a global scale. This integration allows students to transcend geographical boundaries and collaborate with peers from different countries, fostering a greater understanding of diverse cultures and perspectives. By embracing international education technology integration, students can broaden their horizons and develop a global mindset.

One of the primary benefits of international education technology integration is the opportunity for students to engage in virtual exchange programs. Through video conferencing, online forums, and collaborative platforms, students can interact with their counterparts from around the world. This exchange of ideas and experiences enhances cultural awareness, promotes empathy, and encourages a deeper appreciation for global issues.

Furthermore, international education technology integration opens up a world of educational resources and opportunities. Students can access online libraries, virtual museums, and educational platforms that offer courses and materials from renowned institutions worldwide. This wealth of information empowers students to expand their knowledge beyond what traditional classrooms can offer, allowing them to pursue their interests and passions on a global scale.

Moreover, international education technology integration equips students with essential 21st-century skills. By utilizing digital tools and platforms, students develop critical thinking, problem-solving, and digital literacy skills. These skills are highly sought after in today's global job market, where employers value individuals who can navigate and thrive in a technology-driven world.

In conclusion, international education technology integration has a profound impact on students in the field of comparative and international education. It enables students to connect with peers from different countries, access a vast array of educational resources, and develop essential 21st-century skills. By embracing technology as a tool for learning and collaboration, students can break boundaries and prepare themselves for a future that demands a global mindset.

Chapter 2: Understanding Comparative Education Systems

Exploring Different Education Systems around the World

Education is a fundamental right that every student deserves, but the way it is delivered varies greatly across different countries. In our interconnected world, it is crucial for students to understand and appreciate the diverse education systems that exist globally. This subchapter aims to take you on a journey around the world, exploring different education systems and their unique approaches to teaching and learning.

One of the most fascinating aspects of comparative and international education is the opportunity to examine the various educational philosophies that shape different systems. From the progressive pedagogy of Finland to the discipline-focused approach of South Korea, each country has its own educational priorities and values. By delving into these diverse systems, students can gain a deeper understanding of the strengths and weaknesses of their own education system and perhaps find inspiration to implement positive changes.

In this subchapter, we will explore some of the most renowned education systems worldwide. We will delve into the Finnish model, known for its emphasis on creativity, critical thinking, and holistic development. We will also examine the Japanese education system, which values discipline, respect, and diligent work ethic. Furthermore, we will uncover the German education system, renowned for its strong vocational training programs that prepare students for the workforce.

By studying these different systems, students can broaden their perspectives and discover alternative approaches to education. They can learn about the innovative teaching methods, curriculum designs, and assessment strategies employed in different countries. This knowledge can inspire students to think critically about their own education, question the status quo, and advocate for reforms that align with their own values and aspirations.

Additionally, exploring different education systems fosters cultural understanding and appreciation. It allows students to recognize that there is no one-size-fits-all approach to education and that diversity should be celebrated. Learning about the ways in which different countries prioritize different subjects, values, and skills enables students to develop a global mindset and become more open-minded individuals.

In conclusion, this subchapter serves as a window into the world of education systems around the globe. By exploring different models, students can gain insight into the strengths and weaknesses of their own education system, discover innovative teaching practices, and foster cultural understanding. By embracing international education technology integration, students can break boundaries and become empowered agents of change in their own educational journeys.

Comparing Educational Approaches and Methods

In the rapidly evolving landscape of education, it is crucial for students to understand and appreciate the diverse approaches and methods employed worldwide. This subchapter aims to shed light on the various educational approaches and methods utilized across different countries, enabling students to embrace the concept of international education technology integration and broaden their perspectives.

One prominent approach is the traditional lecture-based method, commonly found in many educational systems. This method emphasizes the role of the instructor as the primary source of knowledge, with students passively listening and taking notes. While this approach has its merits, it often lacks student engagement and fails to foster critical thinking and problem-solving skills.

In contrast, the student-centered approach is gaining traction in many progressive educational systems. This method places students at the center of the learning process, encouraging active participation, collaboration, and self-directed learning. Students are empowered to explore their interests, solve real-world problems, and develop a deeper understanding of concepts. This approach promotes critical thinking, creativity, and adaptability, essential skills for the 21st-century workforce.

Another noteworthy method is the project-based learning approach, which encourages students to delve deep into a particular topic through hands-on projects. This approach allows students to apply their knowledge in real-life scenarios, promoting a deeper understanding and retention of concepts. Furthermore, project-based learning fosters teamwork, communication, and problem-solving skills, preparing students for future challenges.

The Montessori method is yet another innovative approach that emphasizes self-paced learning and individualized instruction. Students are encouraged to explore their interests and learn at their own pace, with teachers serving as guides rather than lecturers. This method promotes independence, self-discipline, and a love for lifelong learning.

Furthermore, technology integration has revolutionized education worldwide. The integration of educational technology tools, such as interactive whiteboards, tablets, and online learning platforms, has transformed the traditional classroom into a dynamic and interactive learning environment. This integration enables students to access a wealth of resources, collaborate with peers globally, and engage in personalized and adaptive learning experiences.

By comparing these educational approaches and methods, students can appreciate the strengths and weaknesses of each and gain a broader understanding of the educational landscape. This knowledge will empower them to make informed choices about their own learning journey and embrace the opportunities offered by international education technology integration. By embracing diverse approaches and methods, students can break boundaries and become global citizens equipped with the skills necessary to thrive in an increasingly interconnected world.

Examining Challenges and Successes in Education Systems

Education systems around the world face a multitude of challenges, but they also boast significant successes. In this subchapter, we will explore the complexities and triumphs within various education systems, with a particular focus on the field of Comparative and International Education. By understanding these challenges and successes, students can gain a broader perspective on the global education landscape, and perhaps even find inspiration for innovative solutions.

One of the most pressing challenges in education systems is the issue of access. In many parts of the world, students face barriers in accessing quality education, such as poverty, gender inequality, and inadequate infrastructure. This lack of access hampers the potential of talented individuals and perpetuates social and economic disparities. However, success stories abound, where countries have made significant strides in providing equal educational opportunities for all. By studying these successes, students can learn about effective policies and programs that promote inclusive education.

Another challenge lies in the quality of education. Some education systems struggle with outdated curricula, ineffective teaching methods, and a lack of resources, hindering students' ability to acquire the necessary skills for the 21st century. On the other hand, there are education systems that have successfully implemented innovative teaching practices, tailored curricula, and cutting-edge technologies to enhance learning outcomes. By examining these success stories, students can gain insights into pedagogical approaches that foster critical thinking, creativity, and problem-solving skills.

Furthermore, the role of technology in education cannot be overlooked. While it has the potential to revolutionize learning, its integration into education systems is not without challenges. Issues such as the digital divide, privacy concerns, and the need for teacher training hinder the effective use of technology in classrooms. However, there are education systems that have embraced technology integration, leveraging its power to engage students, facilitate collaboration, and personalize learning experiences. By exploring these successes, students can gain a deeper understanding of the transformative potential of educational technology.

In conclusion, examining challenges and successes in education systems is essential for students interested in Comparative and International Education. By understanding the complexities and triumphs within different education systems, students can develop a global perspective on education and gain insights into innovative solutions. Through studying these challenges and successes, students can contribute to the advancement of education systems worldwide, working towards a more inclusive, equitable, and effective educational landscape.

Chapter 3: The Role of Technology in International Education

Integrating Technology in the Classroom

In today's rapidly evolving world, technology has become an integral part of our daily lives. From communication and entertainment to information and learning, technology has revolutionized the way we interact with the world around us. In the field of education, technology has opened up new avenues for learning and has transformed traditional classrooms into dynamic, interactive spaces. This subchapter explores the importance of integrating technology in the classroom and how it can enhance the educational experience for students in the field of Comparative and International Education.

Technology in the classroom goes beyond using computers and tablets; it encompasses a wide range of tools and resources that can enhance teaching and learning. By integrating technology, students can access information from around the world, collaborate with peers from different countries, and engage in interactive learning experiences that foster critical thinking and problem-solving skills. Technology also enables students to take ownership of their learning, allowing them to explore topics of interest and engage in self-paced learning.

Furthermore, technology integration in the classroom promotes global awareness and prepares students for the increasingly interconnected world they will enter upon graduation. It allows students to connect with their counterparts in different countries, exchange ideas, and gain a deeper understanding of diverse cultures and perspectives. Through virtual exchange programs and online platforms, students can engage in authentic cross-cultural experiences, breaking boundaries and fostering empathy and understanding.

Additionally, integrating technology in the classroom provides access to a wealth of digital resources and online learning platforms. Students can access educational videos, interactive simulations, and online textbooks that cater to their individual learning styles and preferences. This personalized approach to learning ensures that students can grasp complex concepts at their own pace, boosting engagement and academic performance.

However, it is important to recognize that technology integration should not replace traditional teaching methods but rather complement them. Effective integration requires proper planning, professional development for educators, and ongoing support to ensure that technology is used purposefully and effectively.

In conclusion, integrating technology in the classroom is crucial in the field of Comparative and International Education. It opens up endless possibilities for students to explore, collaborate, and learn in a global context. By incorporating technology into their educational journey, students can break boundaries and embrace the opportunities that international education technology integration offers.

Enhancing Learning through Educational Technology

In today's fast-paced world, educational technology has become an integral part of the learning experience. In this subchapter, we will explore how embracing international education technology integration can revolutionize the way students learn and open up new avenues for comparative and international education.

Educational technology refers to the use of digital tools, software, and devices to enhance teaching and learning. It offers students access to a wide range of resources and opportunities that were previously unimaginable. With the click of a button, students can connect with their peers from different parts of the world, access information from renowned institutions, and engage in collaborative projects that transcend geographical boundaries.

One of the key advantages of educational technology is its ability to foster global citizenship and promote intercultural understanding. Through online platforms and virtual classrooms, students can interact with their counterparts from different countries, exchanging ideas and perspectives. This exposure to diverse cultures and viewpoints helps students develop empathy, tolerance, and a broader understanding of the world.

Furthermore, educational technology allows students to access a wealth of educational resources from around the globe. Online libraries, digital textbooks, and interactive learning tools provide students with up-to-date information and engaging content. This access to a wide variety of learning materials enhances the quality of education and promotes self-directed learning.

In the field of comparative and international education, educational technology has the potential to bridge the gap between different education systems. Through online learning platforms and virtual exchange programs, students can compare and contrast educational practices, policies, and curricula from different countries. This comparative approach not only broadens students' knowledge but also helps them critically analyze and evaluate different educational models.

By embracing educational technology, students can also develop essential 21st-century skills such as digital literacy, problem-solving, and collaboration. These skills are highly sought after in today's job market, and by integrating technology into their learning, students gain a competitive edge.

In conclusion, embracing educational technology can transform the way students learn and engage with the world. By incorporating international education technology integration, students can enhance their learning experience, gain a global perspective, and develop essential skills. The possibilities are endless, and the benefits are immense. So, let's embrace educational technology and break boundaries in education!

Overcoming Barriers to Technology Integration in Education

In today's fast-paced and interconnected world, technology has become an essential part of our daily lives. It has transformed the way we communicate, work, and learn. However, when it comes to education, many students still face barriers to fully integrating technology into their learning experiences. This subchapter aims to address these barriers and provide practical strategies for overcoming them.

One of the main barriers to technology integration in education is the lack of access to technology and internet connectivity. In many parts of the world, students do not have access to computers, tablets, or reliable internet connections. This digital divide creates an inequality in educational opportunities, limiting students' access to a vast pool of knowledge and resources. To overcome this barrier, students can explore alternative options such as utilizing public libraries, community centers, or schools that offer technology facilities and internet access. Additionally, advocating for increased government and institutional support for technology infrastructure can help bridge the digital divide and ensure equal access for all students.

Another significant barrier to technology integration is the lack of digital literacy skills among students. Many students are not familiar with using technology for educational purposes, which hinders their ability to effectively utilize digital tools for learning. To overcome this barrier, students can take advantage of various online resources and tutorials that provide step-by-step guidance on using technology for educational purposes. Engaging in online forums and communities can also help students connect with others who have similar interests and can provide support and guidance.

Resistance to change and traditional teaching methods can also pose barriers to technology integration in education. Some students and teachers may be hesitant to embrace technology due to a fear of the unknown or a preference for traditional teaching methods. To overcome this barrier, students can engage in open discussions with teachers and peers, highlighting the benefits of incorporating technology into the learning process. Demonstrating how technology can enhance collaboration, critical thinking, and creativity can help overcome resistance and encourage a more open-minded approach to technology integration.

In conclusion, technology integration in education holds immense potential for enhancing the learning experiences of students. By addressing barriers such as limited access, lack of digital literacy skills, and resistance to change, students can take proactive steps towards embracing technology in their education. By breaking these boundaries, students can fully harness the power of technology to explore new horizons, connect with global communities, and prepare themselves for a future that demands digital fluency.

Chapter 4: Embracing International Education Technology Integration

Fostering Intercultural Communication and Collaboration

In today's interconnected world, the ability to communicate and collaborate with individuals from diverse cultural backgrounds has become increasingly important. As students, you have the opportunity to embrace international education technology integration, allowing you to break boundaries and develop the skills necessary for successful intercultural communication and collaboration.

Intercultural communication refers to the exchange of information and ideas between individuals from different cultural backgrounds. It involves understanding and respecting cultural differences, as well as being able to effectively communicate and collaborate with people from diverse backgrounds. By embracing international education technology integration, you can expand your horizons and gain a deeper understanding of different cultures and perspectives.

One of the key benefits of intercultural communication is the ability to develop empathy and cultural sensitivity. By engaging with individuals from different cultures, you can gain a better understanding of their experiences, beliefs, and values. This increased awareness allows you to become more culturally sensitive and empathetic towards others, fostering a more inclusive and tolerant society.

Moreover, intercultural communication and collaboration are crucial for success in today's globalized workforce. As companies expand their operations internationally, they seek employees who can effectively work with colleagues and clients from different cultural backgrounds.

By developing your intercultural communication and collaboration skills, you will enhance your employability and open doors to exciting career opportunities worldwide.

International education technology integration offers various tools and platforms that can facilitate intercultural communication and collaboration. Virtual exchange programs, online learning platforms, and video conferencing tools enable students to connect with peers from around the world, engage in meaningful discussions, and work on collaborative projects. These technological advancements provide you with the opportunity to interact with individuals from different cultures, share ideas, and learn from one another.

To fully embrace intercultural communication and collaboration, it is essential to approach these interactions with an open mind and a willingness to learn. Embrace diversity, challenge your own assumptions, and actively seek opportunities to engage with individuals from different cultural backgrounds. Remember that intercultural communication is a two-way street, and it is important to listen actively, ask questions, and be respectful of different perspectives.

By fostering intercultural communication and collaboration through international education technology integration, you will not only enhance your personal growth but also contribute to a more interconnected and inclusive world. Embrace the opportunities available to you and break boundaries as you navigate the global landscape.

Developing Global Competencies through Technology

In today's interconnected world, where borders are becoming increasingly porous and communication is instantaneous, global competencies have become essential for students. The ability to understand and appreciate different cultures, communicate effectively across borders, and navigate the complexities of a globalized society is crucial for success in the 21st century. Technology plays a pivotal role in developing these global competencies, enabling students to embrace international education and integrate it seamlessly into their lives.

With the advent of technology, the barriers to accessing global knowledge and connecting with individuals from diverse backgrounds have significantly decreased. Students now have the opportunity to engage in virtual exchanges, collaborate with peers from different countries, and gain firsthand experiences of different cultures without leaving their classrooms. Through video-conferencing, online platforms, and social media, students can engage in cross-cultural dialogues, share perspectives, and develop a global mindset.

One of the key benefits of technology in developing global competencies is the ability to break down language barriers. Translation tools, language learning applications, and online language exchange platforms have made it easier for students to communicate and collaborate with individuals from different linguistic backgrounds. This not only fosters language acquisition but also promotes cultural understanding and empathy.

Furthermore, technology provides students with access to a vast range of resources and information from around the world. Through online databases, e-books, and multimedia platforms, students can explore different educational systems, compare international practices, and gain insights into global issues. Technology also facilitates personalized learning, allowing students to tailor their education to their individual interests and preferences, thus broadening their knowledge and understanding of the world.

In addition to knowledge acquisition, technology enables students to develop crucial skills such as critical thinking, problem-solving, and digital literacy. These skills are essential for navigating the complexities of a globalized society and addressing global challenges. Through online simulations, virtual reality experiences, and collaborative online projects, students can develop their ability to think critically, analyze information from multiple perspectives, and work effectively in diverse teams.

In conclusion, technology has revolutionized the way students engage with international education and develop global competencies. By leveraging the power of technology, students can break boundaries, embrace cultural diversity, and gain the knowledge and skills necessary for success in an interconnected world. Through virtual exchanges, language learning tools, access to global resources, and the development of crucial skills, technology empowers students to become global citizens who can make a positive impact on the world.

Building Digital Literacy Skills for the Global Age

In today's fast-paced and interconnected world, digital literacy skills have become essential for students to thrive in the global age. As technology continues to advance at an unprecedented rate, it is crucial for students to develop the necessary skills to navigate and harness the power of digital tools. This subchapter aims to provide students with valuable insights and strategies to build their digital literacy skills, enabling them to embrace international education technology integration and break boundaries.

Digital literacy encompasses a range of skills, including information literacy, media literacy, and technological literacy. By becoming digitally literate, students can effectively search, evaluate, and use information from various online sources. They can identify credible sources, critically analyze information, and develop their own informed perspectives. In an era of fake news and misinformation, these skills are vital for students to become discerning consumers and creators of content.

Moreover, digital literacy skills enable students to effectively communicate and collaborate with individuals from diverse cultural backgrounds. Technology has transformed the way we connect and interact with others, transcending geographical boundaries. By harnessing digital tools, students can engage in virtual collaboration, sharing ideas and perspectives with peers from around the world. This not only enhances their learning experiences but also fosters global citizenship and intercultural understanding.

To build digital literacy skills, students should engage in hands-on activities that involve the use of technology. They can explore educational websites, participate in online discussions, and create digital content such as videos, blogs, or podcasts. By actively engaging with technology, students develop their technical skills and gain confidence in using digital tools.

Another important aspect of building digital literacy skills is the development of digital citizenship. Students should learn about responsible and ethical online behavior, including issues such as online privacy, cyberbullying, and digital footprints. They should understand the importance of respecting others' opinions and intellectual property, as well as protecting their own digital identities.

Teachers and educational institutions play a crucial role in supporting students' digital literacy development. They can provide guidance and resources, offer training sessions on digital tools and platforms, and create a safe and inclusive digital learning environment. Collaboration between educators and students is essential in fostering digital literacy skills and ensuring that students can fully embrace the opportunities offered by international education technology integration.

In conclusion, building digital literacy skills is essential for students in the global age. By developing these skills, students can effectively navigate the digital landscape, critically analyze information, and communicate with individuals from diverse backgrounds. It is through digital literacy that students can break boundaries and embrace international education technology integration, preparing themselves for the challenges and opportunities of the globalized world.

Chapter 5: Case Studies of International Education Technology Integration

Successful Examples of Technology Integration in Schools

In today's rapidly evolving world, technology has become an integral part of our lives, transforming the way we communicate, work, and learn. As students who are passionate about comparative and international education, it is crucial for us to understand the successful examples of technology integration in schools worldwide. This subchapter aims to inspire and enlighten us about the transformative power of technology in education.

One remarkable example of technology integration is found in South Korea's education system. South Korea has been at the forefront of educational technology, with the government actively promoting digital learning environments. Schools in South Korea have implemented interactive whiteboards, tablets, and virtual reality (VR) tools to enhance student engagement and collaboration. Through these technologies, students can explore virtual worlds, interact with simulations, and access a wealth of information, enabling them to develop critical thinking and problem-solving skills.

Another successful example can be seen in Finland, known for its exceptional education system. Finnish schools have embraced technology to facilitate personalized learning experiences. With the help of online platforms, students can access personalized learning materials and tailor their education according to their interests and learning pace. This approach has resulted in increased student motivation, improved academic performance, and a reduction in achievement gaps.

Moving to the United States, the High Tech High (HTH) network serves as an inspiring model for technology integration. HTH schools integrate technology into their curriculum to foster creativity, collaboration, and innovation. Students are encouraged to use technology as a tool to showcase their learning through multimedia presentations, project-based assignments, and digital portfolios. This approach empowers students to become active creators of knowledge rather than passive recipients.

Furthermore, Rwanda's One Laptop per Child (OLPC) initiative illustrates the transformative impact of technology integration in developing countries. OLPC aims to provide children with laptops, enabling them to access educational content, learn digital skills, and bridge the digital divide. This initiative has revolutionized education in Rwanda, enhancing access to quality education for children in remote areas and empowering them with the necessary tools for a brighter future.

These successful examples highlight the immense potential of technology integration in schools worldwide. By embracing technology, students can experience a more engaging, personalized, and collaborative learning environment. As future leaders in comparative and international education, it is essential for us to advocate for the integration of technology in schools, ensuring that all students have equal opportunities to thrive in the digital age.

Impact of Technology on Student Engagement and Learning Outcomes

In today's rapidly evolving world, technology has become an integral part of our daily lives, affecting almost every aspect of society, including education. The integration of technology in the classroom has revolutionized the way students engage with learning materials and has had a profound impact on their learning outcomes. This subchapter explores the various ways in which technology has transformed student engagement and enhanced learning outcomes in the field of comparative and international education.

One of the key benefits of technology in education is its ability to make learning more interactive and engaging. Students are no longer confined to traditional teaching methods, such as textbooks and lectures. Instead, they have access to a wide range of multimedia resources, including videos, simulations, and interactive learning platforms. These resources not only capture students' attention but also help them visualize complex concepts, making learning more enjoyable and effective.

Furthermore, technology has enabled students to connect and collaborate with peers from different parts of the world. Through online platforms and video conferencing tools, students can engage in virtual discussions, share ideas, and work on projects together. This international collaboration fosters a deeper understanding of cultural diversity and broadens students' perspectives, preparing them for the globalized world they will be entering.

Moreover, technology provides students with personalized learning experiences. Adaptive learning platforms use artificial intelligence algorithms to analyze students' strengths and weaknesses, tailoring the learning materials to their specific needs. This individualized approach enables students to learn at their own pace, ensuring that they grasp the concepts fully before moving on. As a result, students are more likely to achieve higher learning outcomes and develop a deeper understanding of the subject matter.

In conclusion, the impact of technology on student engagement and learning outcomes in the field of comparative and international education is undeniable. Through its interactive nature, technology has transformed the way students engage with learning materials, making the process more enjoyable and effective. Additionally, technology has facilitated international collaboration, exposing students to different cultures and perspectives. Finally, technology has provided personalized learning experiences, enabling students to learn at their own pace and achieve higher learning outcomes. As students in the field of comparative and international education, embracing technology integration is crucial to maximize our learning potential and prepare ourselves for the globalized world.

Lessons Learned from International Education Technology Integration

In today's interconnected world, education technology integration has become a crucial aspect of the learning experience. As students, we have the opportunity to embrace international education technology integration and gain valuable insights that can shape our educational journey and future career prospects. This subchapter aims to highlight some key lessons learned from international education technology integration, with a focus on the comparative and international education niches.

One of the fundamental lessons from international education technology integration is the importance of cultural competence. By utilizing technology in the classroom, we can interact with students and educators from different countries and backgrounds. This exposure enables us to develop a deeper understanding and appreciation for diverse cultures, traditions, and perspectives. We learn to navigate cultural nuances, adapt to different learning styles, and foster a global mindset that prepares us for an increasingly globalized workforce.

Another lesson learned is the power of collaboration. Through education technology integration, we have the opportunity to work on projects and assignments with students from different parts of the world. This collaborative learning environment helps us develop essential skills like teamwork, communication, and problem-solving, all of which are highly valued in the international job market. By working together with peers from diverse backgrounds, we gain a broader perspective on global issues and learn to address challenges through collective effort.

Furthermore, international education technology integration encourages lifelong learning. By engaging with digital resources, online courses, and virtual classrooms, we can access a wealth of knowledge from renowned institutions and experts worldwide. This continuous learning not only enhances our academic abilities but also fosters curiosity, critical thinking, and adaptability – skills that are indispensable in today's fast-paced and ever-evolving world.

Lastly, international education technology integration teaches us the importance of digital literacy and technological proficiency. As we navigate digital platforms, online tools, and virtual learning environments, we become adept at leveraging technology to enhance our learning experience. These skills are highly sought after by employers, as they indicate our ability to adapt to new technologies and work in digital environments, which have become increasingly prevalent in the global job market.

In conclusion, embracing international education technology integration offers students in the comparative and international education niches invaluable lessons. It enhances our cultural competence, fosters collaboration, promotes lifelong learning, and develops digital literacy. By grasping these lessons, we can break boundaries and prepare ourselves for a future where technology and globalization are at the forefront.

Chapter 6: Empowering Students in the Digital Age

Leveraging Technology for Personalized Learning

In today's fast-paced and interconnected world, technology has become an integral part of our daily lives. From communication to entertainment, it has transformed the way we interact and learn. In the field of education, technology has opened up new possibilities for personalized learning, allowing students to tailor their educational experiences to their individual needs and interests. This subchapter explores how students can leverage technology to embrace international education technology integration and break traditional boundaries.

Technology has made learning more accessible and convenient than ever before. With just a few clicks, students can access a wealth of knowledge from around the world. Online platforms, such as Massive Open Online Courses (MOOCs), provide opportunities to learn from renowned educators and experts, regardless of geographical boundaries. By taking advantage of these resources, students can broaden their understanding of different cultures, perspectives, and educational systems, fostering a truly comparative and international education experience.

Moreover, technology allows for personalized learning experiences tailored to individual students' needs. Adaptive learning platforms use algorithms to analyze students' strengths and weaknesses, providing personalized feedback and targeted resources. This enables students to learn at their own pace, focusing on areas where they need the most support and challenging themselves in areas of strength. By leveraging technology, students can take control of their own education, breaking free from the limitations of traditional classroom settings.

Technology also facilitates collaboration and global connections. Through online platforms and social media, students can connect with peers from different countries, engaging in cross-cultural discussions and collaborative projects. This enables them to gain a deeper understanding of international perspectives and develop vital skills for a globalized world. By harnessing technology, students can bridge the gap between comparative and international education, fostering a sense of global citizenship and cultural competence.

In conclusion, technology has revolutionized the field of education, offering students unprecedented opportunities for personalized learning and international education technology integration. By leveraging technology, students can access a world of knowledge and tailor their educational experiences to their individual needs and interests. Through adaptive learning platforms, online collaborations, and global connections, they can break traditional boundaries and embrace a truly comparative and international education. In an increasingly interconnected world, the ability to leverage technology for personalized learning is essential for students seeking to expand their horizons and shape their own educational journeys.

Promoting Creativity and Innovation through Technology

Technology has become an inseparable part of our lives, transforming the way we learn, work, and communicate. In today's globalized world, it is essential for students to embrace international education technology integration to stay ahead and thrive in the competitive job market. This subchapter explores the ways in which technology can promote creativity and innovation among students, focusing on the field of comparative and international education.

One of the key advantages of technology in education is its ability to enhance creativity. With the vast array of digital tools and resources available, students have the opportunity to explore different mediums and express their ideas in unique ways. Whether it's creating multimedia presentations, designing interactive websites, or producing digital artwork, technology empowers students to think outside the box and showcase their creativity to a wider audience.

Furthermore, technology serves as a catalyst for innovation. By leveraging digital platforms, students can collaborate with peers from around the world, exchanging ideas and perspectives. This cross-cultural exchange not only broadens their horizons but also encourages them to think critically and develop innovative solutions to global challenges. Through virtual classrooms, online forums, and social media, students can engage in meaningful discussions and collectively work towards creating a better future.

Moreover, technology provides students with access to a wealth of information and resources. With just a few clicks, they can explore different educational materials, conduct in-depth research, and gain a comprehensive understanding of various subjects. This accessibility enables students to think critically, analyze information, and develop their own unique perspectives. It also encourages them to question existing knowledge and challenge conventional wisdom, fostering a culture of innovation and intellectual growth.

In conclusion, technology is a powerful tool that can promote creativity and innovation among students in the field of comparative and international education. By embracing technology, students can explore new mediums, collaborate with peers worldwide, and access a wealth of information. This integration of technology not only enhances their creativity but also empowers them to become innovative thinkers and global citizens. As students, it is crucial to embrace technology and leverage its potential to break boundaries and make a positive impact on the world.

Preparing Students for Global Careers and Opportunities

In today's interconnected world, the ability to thrive in a global environment is becoming increasingly important. As students, it is crucial to be proactive and well-prepared for the global careers and opportunities that await us. This subchapter aims to guide and equip students with the necessary skills and mindset to navigate the global landscape successfully.

One of the first steps towards preparing for global careers is to foster a global mindset. This means developing an appreciation for different cultures, languages, and perspectives. By embracing diversity and understanding the nuances of various cultures, we can effectively navigate unfamiliar territories and build meaningful connections with people from different backgrounds. Additionally, being open-minded and adaptable are valuable qualities that will enable us to adjust to different work environments and thrive in multicultural teams.

Language proficiency is another crucial aspect of preparing for a global career. Being able to communicate effectively in multiple languages not only enhances our employability but also allows us to connect with people on a deeper level. It is important to invest time and effort in learning languages that are widely spoken in the global arena, such as English, Mandarin, Spanish, or Arabic. Fluency in languages can open doors to exciting job opportunities and enable us to make a positive impact on a global scale.

In addition to language skills, technological literacy is vital for success in the global job market. Technology has become a powerful tool in facilitating global collaboration and communication. Familiarity with various digital platforms, virtual collaboration tools, and social media can greatly enhance our ability to work effectively with individuals from around the world. Moreover, being adept at leveraging technology can help us stay updated with global trends and opportunities.

Furthermore, internships and study abroad programs present invaluable opportunities to gain practical experience and exposure to diverse cultures. These experiences not only allow us to apply classroom knowledge in real-world settings but also foster personal growth and develop our global competencies. Participating in internships or studying abroad can give us a competitive edge in the job market, as employers increasingly value candidates with international experience.

In conclusion, preparing for global careers and opportunities requires a proactive approach and a willingness to embrace diversity. Developing a global mindset, acquiring language proficiency, nurturing technological literacy, and seeking international experience are all essential steps towards building a successful and fulfilling global career. By breaking boundaries and embracing international education technology integration, we can position ourselves for success in the ever-evolving global landscape.

Chapter 7: Overcoming Challenges in International Education Technology Integration

Addressing Infrastructure and Access Issues

In our technologically advanced world, access to proper infrastructure is crucial for students to fully embrace international education and technology integration. However, it is disheartening to acknowledge that many students around the world face significant challenges in this area. This subchapter aims to shed light on the importance of addressing infrastructure and access issues and provide potential solutions for students, particularly within the field of Comparative and International Education.

The digital divide remains a major concern when it comes to accessing educational resources. Many students in developing countries lack access to reliable internet connections, computers, or even electricity. This hinders their ability to engage with international educational platforms and limits their opportunities for collaboration and knowledge exchange. To bridge this gap, governments and educational institutions must prioritize investing in infrastructure development, ensuring that all students have equal access to technological resources.

One possible solution is the establishment of digital learning centers or hubs in underserved communities. These centers could provide students with access to computers, internet connections, and other necessary equipment. Additionally, initiatives such as providing students with affordable or subsidized laptops or tablets could greatly enhance their ability to utilize educational technology.

Collaboration between governments, non-profit organizations, and the private sector is crucial in addressing infrastructure and access issues. Through partnerships, these entities can work together to develop innovative solutions and implement sustainable infrastructure projects. For instance, telecommunications companies could invest in expanding internet coverage to remote areas, while governments can offer incentives to attract private investments in educational technology.

Furthermore, it is essential to prioritize training and capacity-building programs for teachers and students. Educators should be equipped with the necessary skills to effectively integrate technology into their teaching practices, while students should receive training on using digital tools for learning. By investing in professional development and digital literacy programs, we can ensure that students are not only provided with access to technology but also empowered to use it effectively.

In conclusion, addressing infrastructure and access issues is vital for students to fully embrace international education and technology integration. By investing in infrastructure development, establishing digital learning centers, fostering partnerships, and providing training opportunities, we can bridge the digital divide and create a more inclusive educational environment. It is imperative that governments, educational institutions, and stakeholders prioritize these efforts to ensure equal access to educational opportunities for all students in the field of Comparative and International Education.

Training and Professional Development for Educators

In today's rapidly evolving world, education has become a global concern. As students, it is crucial for us to understand the importance of training and professional development for educators in the realm of comparative and international education. This subchapter delves into the significance of equipping teachers with the necessary skills and knowledge to embrace international education technology integration.

Effective teaching is not limited to imparting knowledge from textbooks; it extends beyond the boundaries of traditional methods. To prepare educators for the challenges of a globalized world, training and professional development play a pivotal role. By staying updated with the latest advancements in technology and teaching methodologies, educators can ensure that their students receive a comprehensive and well-rounded education.

One key aspect of training for educators is the integration of international education technology. With the rise of digital platforms and online learning, it has become essential for teachers to incorporate technology into their classrooms. Through training programs, educators can learn to utilize various educational technologies, such as virtual reality, online collaboration tools, and interactive multimedia resources. These tools not only enhance the learning experience but also foster global awareness and cultural understanding among students.

Furthermore, professional development programs for educators in comparative and international education focus on developing intercultural competencies. Teachers learn to navigate cultural differences and adapt their teaching strategies accordingly. By understanding diverse perspectives and cultures, educators can create inclusive learning environments that promote tolerance and mutual respect.

Training and professional development also provide educators with opportunities for collaboration and networking. Workshops and conferences bring together educators from different backgrounds, allowing them to exchange ideas and best practices. This collaboration fosters a sense of community and inspires educators to constantly improve their teaching methods.

As students, we benefit greatly from well-trained and globally aware educators. Their expertise in international education technology integration ensures that we receive an education that prepares us for a globalized world. By embracing technology and understanding diverse cultures, educators equip us with the necessary skills to thrive in an interconnected society.

In conclusion, training and professional development for educators in comparative and international education are imperative for fostering effective teaching practices. By integrating international education technology and developing intercultural competencies, educators can provide us with a well-rounded education. As students, we appreciate and benefit from the dedication of educators who continuously strive to expand their knowledge and skills in order to enhance our learning experience.

Navigating Ethical and Privacy Concerns in Technology Integration

In today's rapidly evolving digital landscape, technology integration has become an essential part of education. It offers students access to a wealth of information, enhances collaboration, and encourages critical thinking. However, as we embrace the benefits of technology integration, it is crucial to address the ethical and privacy concerns that come with it.

One of the primary ethical concerns in technology integration is the issue of data privacy. When using digital tools and platforms, students often share personal information, such as their names, contact details, or even sensitive data like health records. It is essential to understand how this data is collected, stored, and used. Students should be aware of their rights and be cautious about sharing personal information online. Educators can play a vital role in teaching students about data privacy, encouraging them to think critically before sharing any personal information.

Another ethical concern is the digital divide. While technology integration offers many opportunities, it also exacerbates the gap between students who have access to technology and those who do not. It is crucial to ensure that all students, regardless of their socio-economic background, have equal access to technology. Schools and policymakers need to work together to bridge this divide by providing resources, such as computers and internet access, to underserved communities.

Furthermore, technology integration raises concerns about digital citizenship and online behavior. Students must understand the importance of responsible digital citizenship, including respectful online communication, protecting their online identities, and avoiding cyberbullying. Educators can foster discussions and workshops on digital citizenship to help students develop the necessary skills to navigate the online world ethically.

Lastly, there is an ethical and moral responsibility to use technology in a way that promotes inclusivity and diversity. We must be mindful of the potential biases and discrimination that can arise in technology integration. Students should be educated on the ethical considerations related to algorithmic bias, AI, and machine learning. By understanding these issues, students can actively work towards creating a more inclusive and fair digital environment.

In conclusion, while technology integration offers numerous benefits in education, it is crucial for students to navigate the ethical and privacy concerns that accompany it. Data privacy, the digital divide, responsible digital citizenship, and promoting inclusivity are all important aspects to consider. By addressing these concerns and fostering ethical practices, students can embrace technology integration while ensuring a safe and inclusive digital environment.

Chapter 8: Future Trends in International Education Technology Integration

Emerging Technologies in Education

In today's rapidly evolving world, the integration of technology in education has become a necessity rather than a luxury. As students, you are at the forefront of this digital revolution, embracing international education technology integration. This subchapter, "Emerging Technologies in Education," will explore the latest technological advancements that are transforming the way we learn, collaborate, and engage with educational content globally.

One of the most significant emerging technologies in education is virtual reality (VR). Imagine being able to explore ancient civilizations, dive into the depths of the ocean, or even travel to outer space without leaving your classroom. VR provides immersive experiences that enhance your understanding of complex subjects and bring learning to life. Through virtual field trips, interactive simulations, and 3D modeling, you can now explore the world like never before.

Another game-changing technology is augmented reality (AR). Unlike VR, which creates an entirely virtual environment, AR overlays digital information onto the real world. This technology enables you to interact with virtual objects and characters in real-time, making learning interactive and engaging. For example, you can dissect virtual organs, solve math problems by visualizing equations in 3D, or explore historical events through interactive timelines.

Artificial intelligence (AI) is also revolutionizing education. AI-powered tools can personalize learning experiences by analyzing data and adapting content to individual students' needs. With AI, you can receive immediate feedback, access personalized recommendations, and engage in adaptive learning experiences that cater to your unique learning style.

Furthermore, the internet of things (IoT) is transforming classrooms into smart learning environments. IoT devices, such as smartboards, connected devices, and wearable technology, provide real-time data that enables personalized instruction, efficient resource management, and collaborative learning. These devices enhance communication, foster collaboration, and promote critical thinking skills essential for the 21st-century workforce.

In conclusion, emerging technologies in education are reshaping the way we learn, collaborate, and engage with educational content. Virtual reality, augmented reality, artificial intelligence, and the internet of things are just a few examples of the exciting advancements that offer limitless possibilities for students like you. By embracing these technologies, you can broaden your horizons, enhance your learning experiences, and prepare yourselves for a future where technology is an integral part of every aspect of our lives. Embrace the digital revolution and unlock your full potential as global learners.

The Role of Artificial Intelligence in Education

In the ever-evolving world of technology, artificial intelligence (AI) has emerged as a powerful tool with the potential to revolutionize the field of education. As students in the era of international education technology integration, it is crucial for you to understand the role AI plays in shaping the future of learning.

AI has the capability to personalize education like never before. By analyzing vast amounts of data, AI algorithms can identify each student's unique learning style, strengths, and weaknesses. This allows for personalized lesson plans, adaptive assessments, and tailored feedback, ensuring that students receive a customized education experience that caters to their individual needs. No longer will you be confined to a one-size-fits-all approach; AI empowers you to learn at your own pace, fostering a deeper understanding of the subject matter.

Moreover, AI can provide students with access to a wealth of educational resources. Through intelligent tutoring systems and virtual assistants, AI can act as a guide, facilitating learning beyond the confines of the classroom. Whether it is answering your questions, suggesting additional reading materials, or even providing real-time language translation, AI has the potential to enhance your educational journey and break down barriers to learning.

Another significant aspect of AI in education is its ability to automate administrative tasks. Grading papers, organizing schedules, and managing student records can be time-consuming for educators. With AI, these mundane tasks can be automated, freeing up valuable time for teachers to focus on what truly matters – teaching and mentoring students. This not only improves efficiency but also enables educators to provide more personalized attention to their students, fostering stronger teacher-student relationships.

While AI undoubtedly offers numerous benefits, it is important to approach its integration in education with caution. Ethical considerations, such as data privacy and algorithm bias, must be carefully addressed to ensure the responsible and equitable use of AI in the classroom. As future global citizens, it is your responsibility to be aware of these challenges and actively participate in shaping the ethical framework surrounding AI in education.

In conclusion, the role of artificial intelligence in education is transformative. From personalized learning experiences to access to vast educational resources, AI has the potential to revolutionize how we learn and teach. As students in the field of comparative and international education, it is crucial for you to embrace AI as a powerful tool that can bridge gaps and break boundaries, fostering a truly inclusive and equitable education system.

Envisioning the Future of International Education Technology Integration

As students, you are living in a world that is becoming increasingly interconnected and globalized. The rapid advancement of technology has transformed the way we learn and communicate, breaking down the barriers of time and space. In this subchapter, we will explore the future of international education technology integration and its implications for students like you.

Education technology integration refers to the use of digital tools and resources to enhance the learning experience and facilitate global collaboration. It enables students to connect with peers from different cultures, gain a deeper understanding of global issues, and develop essential skills for the 21st century.

The future of international education technology integration holds immense potential. Imagine a classroom where students can interact with their counterparts from across the globe, sharing ideas, collaborating on projects, and learning from diverse perspectives. Technology enables real-time communication, breaking down geographical boundaries and fostering a sense of global citizenship.

Furthermore, virtual reality (VR) and augmented reality (AR) technologies are revolutionizing the way we learn. Imagine stepping into a virtual classroom where you can explore historical landmarks, visit museums, or even travel to different countries without leaving your seat. These immersive experiences not only make learning engaging and interactive but also provide a deeper understanding of different cultures and societies.

Artificial intelligence (AI) is another technology that is set to transform education. AI-powered virtual tutors can provide personalized learning experiences, catering to individual students' needs and pace of learning. Adaptive learning platforms can track your progress, identify areas of improvement, and recommend tailored resources to enhance your learning journey.

As students pursuing studies in comparative and international education, embracing technology integration is crucial. It equips you with the skills and knowledge needed to thrive in an increasingly globalized world. By leveraging education technology, you can connect with students from diverse backgrounds, gain a deeper understanding of global issues, and develop cross-cultural competency.

However, it is essential to ensure that technology integration is done responsibly and ethically. As technology advances, it is crucial to address issues such as privacy, security, and the digital divide. Access to technology and digital literacy should be made available to all students, regardless of their socioeconomic background.

In conclusion, the future of international education technology integration holds immense promise for students like you. It provides opportunities to connect, collaborate, and learn from peers across the globe. By embracing technology, you can develop essential skills for the 21st century, gain a global perspective, and become active global citizens. However, it is important to approach technology integration responsibly and ensure equitable access for all.

Chapter 9: Conclusion

Recap of Key Findings and Takeaways

In the captivating book, "Breaking Boundaries: Students Embrace International Education Technology Integration," the authors delve into the world of comparative and international education, shedding light on the transformative power of integrating technology in the classroom. As students ourselves, it is crucial for us to understand the key findings and takeaways from this insightful read.

Throughout the book, the authors present a wealth of evidence showcasing the positive impact of international education technology integration. They highlight how technology has become a bridge connecting students from different cultures, fostering global understanding and collaboration.

One of the key findings emphasized in the book is the immense potential of technology in breaking down geographical barriers. Through virtual exchange programs and online collaborative projects, students have the opportunity to interact with peers from diverse backgrounds, gaining a deeper appreciation for different cultures and perspectives. This exposure helps to cultivate empathy, foster intercultural skills, and prepare students to thrive in an increasingly globalized world.

Furthermore, the book emphasizes the importance of critical thinking and problem-solving skills in today's interconnected society. By incorporating technology into the curriculum, educators can provide students with authentic, real-world learning experiences. From conducting research using digital resources to analyzing data and creating multimedia presentations, students develop crucial skills that go beyond rote memorization. These skills are essential for success in the modern job market, where adaptability and creativity are highly valued.

Additionally, the authors discuss the role of technology in promoting student engagement and motivation. Through interactive learning platforms, gamification, and multimedia resources, educators can create a dynamic and stimulating learning environment. Students become active participants in their education, taking ownership of their learning journey and developing a lifelong love for learning.

In conclusion, "Breaking Boundaries: Students Embrace International Education Technology Integration" offers valuable insights into the world of comparative and international education. The book highlights the transformative power of technology in breaking down barriers, fostering global understanding, and cultivating essential skills for success in the 21st century. As students, we must take note of these key findings and embrace the opportunities offered by technology to broaden our horizons and become global citizens.

Empowering Students to Embrace International Education Technology Integration

In today's interconnected world, the integration of technology in education has become an essential component of preparing students for the global workforce. With the rapid advancements in communication and information technology, students now have the opportunity to connect and collaborate with peers from all around the world, breaking down the boundaries that once limited their educational experiences.

This subchapter aims to empower students to embrace international education technology integration and harness its potential for their own educational growth. By embracing technology, students can expand their horizons, gain global perspectives, and develop skills that are crucial in the field of comparative and international education.

One of the key benefits of international education technology integration is the ability to connect with students from different cultural backgrounds. Through online platforms, students can engage in virtual classrooms, participate in collaborative projects, and exchange ideas with peers from different countries. This exposure to diverse perspectives fosters empathy, cultural understanding, and promotes a global mindset.

Furthermore, technology integration allows students to access a wealth of educational resources from around the world. Online libraries, digital textbooks, and interactive learning platforms provide students with a vast array of information at their fingertips. This not only enhances their knowledge base but also encourages independent learning and critical thinking skills.

Moreover, technology integration facilitates the development of digital literacy skills that are highly valued in the field of comparative and international education. Students learn to navigate digital tools, analyze and evaluate online information, and effectively communicate their ideas. These skills are transferable across various disciplines and are essential in today's digital age.

To fully embrace international education technology integration, students are encouraged to actively seek out opportunities for international collaboration. They can join virtual exchange programs, participate in online discussions, and engage in global projects. These experiences not only enrich their educational journey but also provide them with a competitive edge in the job market.

In conclusion, embracing international education technology integration opens up a world of opportunities for students in the field of comparative and international education. By connecting with peers from different cultural backgrounds, accessing global educational resources, and developing digital literacy skills, students are empowered to excel in an increasingly interconnected world. Embracing technology integration is not only beneficial for their educational growth but also crucial for their future success.